Experimenting
WITH

Even with this lever, I can't move this rock. What now?

I guess you'll have to LEVER right there!

Simple Machines

Gordon R. Gore

Trifolium Books Inc.
Toronto, Canada

Trifolium Books Inc.
250 Merton Street, Suite 203
Toronto, Ontario, Canada M4S 1B1
Tel: 416-483-7211 Fax: 416-483-3533
e-mail: trifoliu@ican.net

We acknowledge the financial support of the Government of Canada through the Book Publishing Industry Development Program (BPIDP) for our publishing activities.

Canadä

Canadian Cataloguing in Publication Data

Gore, Gordon R.
   Experimenting with simple machines: hands-on science activities

ISBN 1-55244-038-9

1. Simple machines—Experiments—Juvenile literature. I. Title. II. Title: Simple machines.
TJ147.G67 2000   621.8'078   C00-930551-3

Please note: In most cases, the terms and units in this book have been presented in standard S.I. style. In some instances, however, the author has deviated from S.I. in order to clarify ideas, reinforce new terms, and/or improve the flow of the text.

Project editor: Sara Goodchild
Design and layout: Heidy Lawrance Associates
Project coordinator: Sara Goodchild
Production coordinator: Heidy Lawrance Associates
Cover design: Heidy Lawrance Associates
Cartoons: Ehren Stillman; Computer line drawings: Moira Rockwell

Printed and bound in Canada
10 9 8 7 6 5 4 3 2 1

Trifolium's books may be purchased in bulk for educational, business, or promotional use. For information, please contact: Special Sales, Trifolium Books Inc., 250 Merton Street, Suite 203, Toronto, Ontario, M4S 1B1 Tel: 416-483-7211 Fax: 416-483-3533.

Introduction

Experimenting with Simple Machines is a book of hands-on science activities for young people, dealing with levers, pulleys, inclined planes, wheels-and-axles and other simple machines students use every day (whether they realize it or not).

The activities require minimal equipment. The experiments will, however, be enriched if students can measure forces with newton scales, which measure force in international metric system units.

The unit will be especially valuable if the teacher can gather a collection of everyday simple machines, and relate these real items to the descriptions in the book. Some everyday items that would be useful include: scissors, a clothesline pulley, a stapler, a staple remover, a can opener, a bottle opener, a pencil sharpener, a crow bar, a hammer, a screwdriver, pliers, a wrench, and so on. (Even a baseball bat is a simple machine!)

This unit is a very practical one, and can be made more practical if students are encouraged to relate what they are learning to the real world. A constant theme might be "How could we improve on this device (tool) so it might be even easier or more convenient to use?"

When students have finished this unit, they might be encouraged to make a model machine *system*, using scrap materials. Along with their model, they might prepare a chart describing the parts of their model and what kind of simple machines make up their 'system'. Some students might enjoy creating a 'Rube Goldberg' device, which uses a chain of simple machines to accomplish some light-hearted chore.

In the author's view, the main objective of an elementary science unit is to build on the natural interest students have in their physical world, by providing them with as much concrete, truly hands-on experience as possible. Hands-on science makes the subject real and enjoyable.

The author hopes this book will be used to let students do some science and learn about the basics of simple machines through direct experience. If, when they finish using this book, students are eager to learn more about simple machines or science in general, then my main objective for elementary science will have been achieved.

Acknowledgments

Thank you to my favorite cartoonist Ehren Stillman, of Abbotsford, B.C., for contributing the cartoons that appear in this book. Ehren's work has been published in many science and physics books by this author, in two chemistry texts by Dr. Jim Hebden, in *The Physics Teacher* (New York) and in *The Science Teacher* (Washington, D.C.). Most of the computer line drawings in this book and in the teacher's guide are by Moira Rockwell of Mission, B.C. Her fine work is always greatly appreciated. All photographs are by the author.

Safety

Introduction to Safety in the Science Lab

The activities in this book have been teacher and student-tested, and are safe when carried out as directed, with proper care. These activities should not be attempted without the permission and supervision of a teacher.

When special attention to safety is required, an icon like this ⚠ appears beside the activity.

In addition to the routine safety procedures used in your school, there are some general guidelines that are important to follow when working through any science activity.

Safety Guidelines for Students

When you begin

- Listen carefully to your teacher's instructions.
- Make sure your teacher knows if you have allergies or any other medical or physical conditions that might affect your ability to carry out activities safely.
- If you wear a hearing aid or contact lenses, tell your teacher before starting any activities.
- Start working on activities only after your teacher tells you to begin. Get your teacher's permission before changing any steps in an activity.
- Know the location and proper operation of blanket, fire extinguisher, fire alarm and any other emergency equipment.

In the Lab

- Wear a lab apron, safety goggles, and/or closed-toe shoes when your teacher directs you to do so.
- Make sure long hair or loose clothing is tied back.
- Never eat or drink while you are working.
- Work quietly and carefully with a partner, and pay attention to the actions of others around you as well as your own.
- Inform your teacher immediately of any damaged equipment or glassware, any accident, or any behavior that you consider dangerous.

Experiments using Electricity

- Your work area and your hands should be dry when in contact with any electrical equipment.
- Always unplug electrical equipment by pulling the plug, not the cord.
- Report any damaged equipment to your teacher.
- Make sure electrical cords are out of the way so that you or others will not trip over them.

Experiments using Heat

- Always use heatproof containers that are whole and undamaged.
- Never use a Bunsen burner or hot plate without the permission and supervision of your teacher.
- Direct the open end of a container that you are heating away from yourself and your classmates.
- Never allow containers to boil dry.
- If you burn yourself, inform your teacher, and cover the burned area with cold water or ice immediately.

Experiments using Chemicals

- Handle substances carefully, and do not touch substances unless you are instructed to do so.
- If you are instructed to smell a substance, never smell it directly. Instead, hold the container slightly away from your face and waft fumes towards your nose with your hands.
- Always hold containers away from your face when pouring liquids.
- If any part of your body touches a substance, rinse the area immediately and continuously with water. If something gets into your eyes, do not touch them. Immediately rinse them with water for fifteen minutes, and inform your teacher.

Cleanup

- Wash your hands after completing any activity.
- Clean all equipment and put it away according to your teacher's instructions.
- Follow your teacher's instructions for cleaning up spills and disposing of materials.

Table of Contents

SECTION 1 • Levers

The one simple machine that you use most is the **lever**. Everyone uses levers, whether they know it or not. Your arms, legs and jaw are levers. Baseball bats, tennis rackets, hockey sticks and golf clubs are levers. Scissors, bottle openers, chemical balances, seesaws, wrenches, crowbars and hammers are also levers.

Why do you use levers? Sometimes you use them because they reduce the **effort force** needed to move a load. In other words, you may use levers to gain a **force advantage**. But you may also use levers to obtain a speed advantage. For example, when you use a broom to sweep a floor, or swing a baseball bat to hit a ball, you get a **speed advantage**.

A lever is often a bar of some sort, with a pivot around which the bar can rotate. The pivot is called the **fulcrum**. See **Figure 1**.

First Class Levers

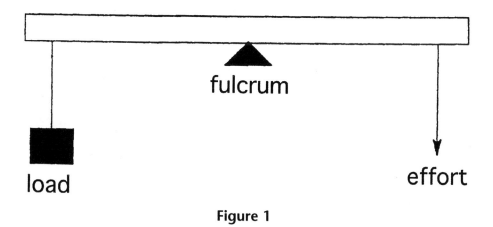

Figure 1

In a **first class lever (Figure 1)**, the **fulcrum** is located between the **load** and the **effort force**. A seesaw **(Figure 3)** is one good example of a first class lever.

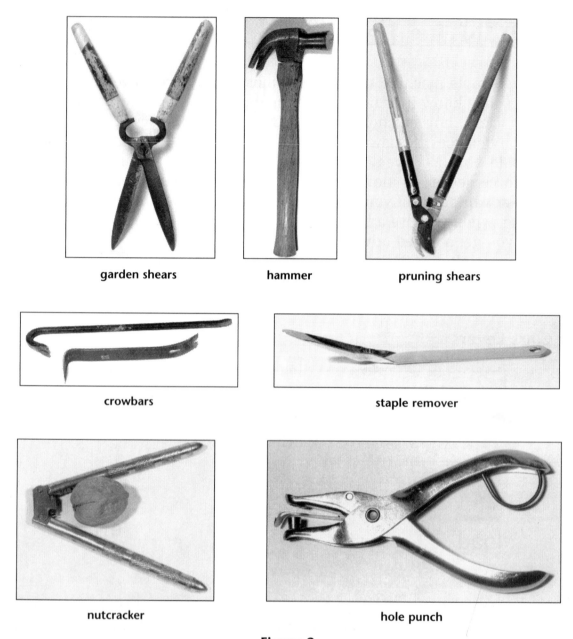

garden shears   hammer   pruning shears

crowbars   staple remover

nutcracker   hole punch

Figure 2

All of the devices in **Figure 2** are **levers**. A lever is one kind of **simple machine**. You use simple machines for several reasons. Sometimes you use a simple machine because it makes it easier for you to lift a heavy load or to overcome a strong force with a weaker force. Sometimes you use a simple machine because it allows you to move something faster than you could move it without the machine.

EXERCISE CARE
WHEN USING
SHARP EQUIPMENT

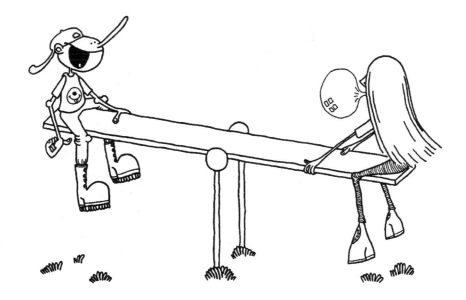

Figure 3 A seesaw is an example of a first class lever.

Second Class Levers

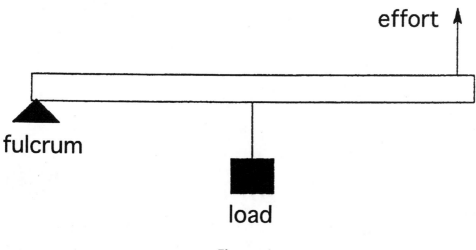

effort

fulcrum

load

Figure 4

A **second class lever** has the fulcrum at one end. (See **Figure 4**). The **effort force** is exerted at the other end. The **load** is somewhere in between the fulcrum and the effort force. A wheelbarrow and a can opener (**Figures 5** and **6**) are both examples of second class levers.

Figure 5

Figure 6

A **wheelbarrow (Figure 5)** and a **can opener (Figure 6)** are good examples of second class levers. On a wheelbarrow, the axle of the wheel is the fulcrum. You apply your effort force at the end of the handles. The load is placed as close to the fulcrum as possible. Where are the fulcrum, the load and the effort on the can opener?

EXERCISE CARE
WHEN USING
SHARP EQUIPMENT

Third Class Levers

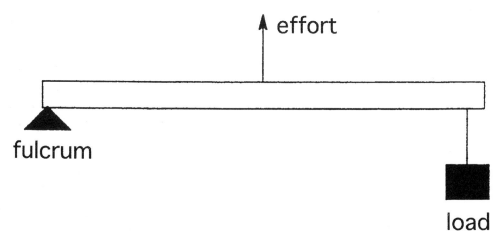

Figure 7

A **third class lever** (**Figure 7**) has the **fulcrum** at one end and the **load** at the other. The **effort force** is applied anywhere between the fulcrum and the load. Third class levers are very common. Baseball bats, golf clubs, hockey sticks, brooms, fishing rods, your arms, your legs and your jaw are all third class levers.

1.1 Try This Yourself! Three Kinds of Levers

What You Need
1 metre stick
1 fulcrum (pivot) for the metre stick
a kilogram mass (or similar load)
thin string or wire to attach the load to the metre stick

What To Do

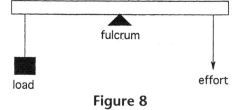

Figure 8

1. Set up a metre stick so that the fulcrum is midway between a kilogram mass (load) and your effort force, as in **Figure 8**. This is a first class lever.

2. Experiment with different positions of the load, fulcrum and effort force, and find out which arrangement lets you lift the load with the least effort force.

3. Sketch the arrangement that allows you to use the least effort force. On your sketch, show the fulcrum, load and effort force.

4. Set up a metre stick as a second class lever, as in **Figure 9**. The fulcrum is at one end of the metre stick, and the load is midway between the fulcrum and the other end of the metre stick.

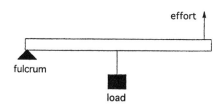

Figure 9

5. Experiment to see what arrangement of fulcrum, load and effort lets you lift the load with the the least effort force.

6. Sketch the arrangement that allows you to use the least effort force.

7. Set up a third class lever as in **Figure 10**. You will have to hold down the metre stick at the pivot. *Imagine the lever is a fishing rod, and the load is a fish.*

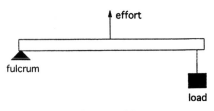

Figure 10

8. Describe what is it like trying to lift the 'fish' by applying an effort force anywhere near the fulcrum.

Some Common Levers

human arm	seesaw
equal-arm balance	hammer
wrench	human jaw
automobile jack	can piercer
scissors	crow bar
wheelbarrow	stapler
nut cracker	paper cutter
tennis racket	bottle cap remover
hockey stick	bicycle pedals
broom	handlebars
golf club	fishing rod
nail clipper handle	human leg

Questions

1. A lever is often used to make it easier to lift a heavy load. If a lever reduces the effort force you need to move a load, we say the lever provides a **force advantage**.

   From the list of common levers above, choose at least five devices that are used to provide a force advantage.

2. Many levers are used to to obtain a **speed advantage** rather than a force advantage. Third class levers are used for this reason.

   From the list of common levers above, choose at least five devices that provide a speed advantage.

Measuring Forces

Scientists measure forces in units called **newtons (N)**. For example, the force of gravity on a one-kilogram mass is 9.8 N, or almost 10 N. Spring scales used in a science laboratory will be marked in newtons.

If you multiply your own mass in kilograms by ten, this will tell you your approximate weight in newtons.

Mechanical Advantage

If you divide the actual load lifted by a simple machine by the **effort force** *you* have to exert to lift it, you obtain a number called the **mechanical advantage** of the machine. If the load is greater than the effort, the machine has a mechanical advantage greater than one. This machine provides a **force advantage**. If the effort is greater than the load, the machine has a mechanical advantage less than one. This machine provides a **speed advantage**.

Example

A student uses a lever to lift a 20 newton load, with an effort of only 10 newtons. What is the mechanical advantage of the lever?

$$\text{Mechanical Advantage} = \frac{\text{load}}{\text{effort force}} = \frac{20 \text{ N}}{10 \text{ N}} = 2$$

Questions

1. **(a)** Using a pulley system, you might lift a 500 N load using an effort force of only 100 N. What is the mechanical advantage of the pulley system?

   (b) Does this pulley system provide a force advantage or a speed advantage?

2. **(a)** Using a fishing rod, an angler requires an effort force of 100 N to lift a 20 N fish out of the water. What is the mechanical advantage of the fishing rod?

   (b) Does this fishing rod provide a force advantage or a speed advantage?

Torque

What do door keys, steering wheels, screwdrivers, pliers and wrenches all have in common? All of these simple machines make it easier for you to make something *rotate* about an axle.

When you push a door open, you apply your force as far from the hinges of the door as possible, because this gives you the greatest turning effect. The *turning effect* of a force is called its torque.

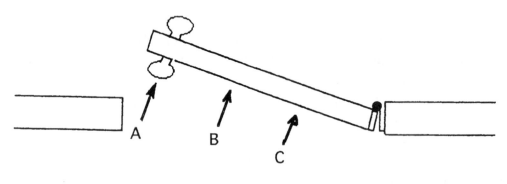

Figure 11

In **Figure 11**, how much **torque** you apply to a door depends on two things:

(1) how hard you push (how much force you exert), and
(2) how far from the hinge you apply the force.

The distance from the hinge is called the **lever arm**. The amount of torque you apply depends on both the force you apply and the lever arm.

1.2 Try This Yourself! Torque to Your Door

It requires a certain amount of torque to open a door. Torque depends on the force you apply and on where you apply it.

1. Try pushing a door open. First, push it open the normal way. Apply the force (push) at or near the door handle (**A**).

2. Now push the same door open by applying your force at **B**, which is closer to the fulcrum (hinge) of the door.

3. Try opening the door by pushing it very near the fulcrum, at **C**.

Questions

1. Is the **effort force** you need greatest at **A**, **B** or **C**?

2. Is the **lever arm** greatest at **A**, **B** or **C**?

3. To make the effort force as small as possible, what must you do to the lever arm?

4. Why is a screwdriver easier to use if it has a fatter handle?

5. Name at least three common tools you use that are made specifically to provide extra **torque** (to make it easier to turn something).

Challenge

Design an experiment to actually measure the force needed to open a door at a slow, steady speed, when the force is applied at several measured distances from the hinge. Calculate the torque needed in each trial.

1.3 Try This Yourself! Balancing Torques

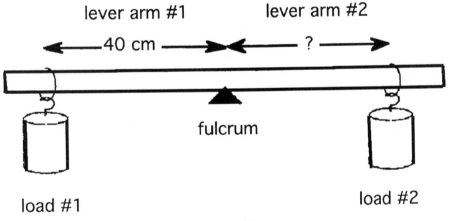

Figure 12

What You Need
1 metre stick
1 pivot for the metre stick
1 smaller load (Example: a bundle of 10 identical washers)
1 larger load (Example: a bundle of 20 identical washers)
2 thin wires or strings to attach loads to a metre stick

What to Do

1. Balance a metre stick on a fulcrum (pivot), so that it does not rotate one way or the other. The pivot should be very close to the 50 cm mark on the metre stick.

2. Hang the smaller load (for example, a bundle of 10 washers) 40 cm from the fulcrum, using a light string. The lever arm is 40 cm.

3. Balance the metre stick again by adding the larger load (for example, a bundle of 20 washers) on the other side of the fulcrum. Measure the distance from the pivot to the larger load, and write it down. This is the lever arm for the other side of the metre stick.

4. **(a)** On the left side of the pivot, multiply the load (number of washers), by the lever arm, in centimetres. Write down the product.

    load x lever arm on left side = _____ x _____ = _____.

    (b) On the right side of the pivot, multiply the **load** (number of washers) by the lever arm, in centimetres. Write down the product.

    load x lever arm on right side = _____ x _____ = _____.

    (c) Compare the two products. Are they almost the same?

5. Try balancing the metre stick with a different combination of loads and lever arms. When the metre stick is balanced, multiply **load x lever arm** on each side of the pivot. Write down these products.

    load x lever arm on left side = _____ x _____ = _____.

    load x lever arm on right side = _____ x _____ = _____.

Question

When the metre stick is balanced, the loads on each side may be different, and the lever arms may be different. What is it that is equal (balanced) on both sides of the pivot?

What is Torque?

Torque is the product of a force and the distance from where it is applied to the pivot. In other words, torque is equal to force multiplied by lever arm.

Torque = Force x Lever Arm

In the previous experiment, you may have found that the torque on one side of the balanced metre stick was equal to the torque on the other side.

Figure 13

Project

Make a **mobile**, using scrap materials and a variety of interesting loads. See **Figure 13**. If the mobile is balanced, what must be true of the various torques acting on it?

1.4 Try This Yourself! Centre Of Gravity

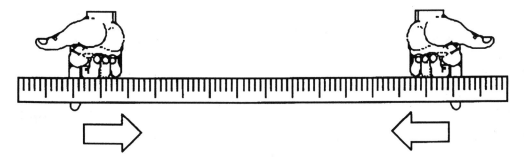

Figure 14

You can find the **centre of gravity** or 'balance point' of a metre stick easily. Hold the metre stick so that one end of it is on your left index finger and the other end is on your right index finger, as in **Figure 14**. Slide your fingers toward each other slowly. They will meet at the 'balance point' every time! The balance point is at the metre stick's centre of gravity.

When you are supporting an object like this metre stick at its **centre of gravity**, the object experiences no torque. (It does not tend to rotate one way or the other.)

If you try to support a metre stick by placing it on a fulcrum which is not at the centre of gravity, the metre stick will no longer balance. It will experience a torque now, as if the force of gravity on the metre stick acted through the **centre of gravity**!

SECTION 2 • Wheels-and-Axles

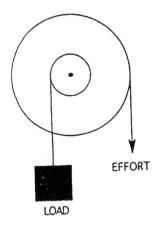

EFFORT

LOAD

Figure 15

The wheel-and-axle (**Figure 15**) is another very common type of simple machine. It is really just a lever that can rotate one full circle. You probably use several wheels-and-axles during a typical day. See **Figure 16**.

Doorknobs, keys, screwdrivers, water taps, and pencil sharpener handles are all wheels-and-axles. Meat grinders, fishing reels and steering wheels are also wheels-and-axles.

2.1 Try This Yourself! Wheels-and-Axles

What You Need
1 piece of scrap wood, about 5 cm x 10 cm x 30 cm
2 wood screws, about 2.5 cm long
1 screwdriver to match the type of screw used.

HANDLE SCREWS AND SCREWDRIVERS WITH CARE

What to Do

1. Using the screwdriver in the normal way, drive one screw into the scrap wood.

2. Try driving the second screw into the wood but, this time, do not use the handle. Use only the metal stem of the screwdriver.

steering wheel

faucet

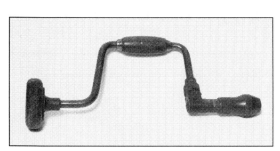

hand drill

pencil sharpener

door key

key in a wind-up toy

Figure 16

Questions

1. Why is it so much easier to drive a screw into wood when you use the handle of the screwdriver? Use the terms torque, lever arm and effort force in your answer.

2. Sketch a screwdriver handle as seen from the end of the handle (see **Figure 17**), and show on the diagram where the force is applied, where the pivot is, and where the lever arm is.

EXERCISE CARE
WHEN USING
EQUIPMENT

3. How would you change the design of the screwdriver to make it even easier to drive a screw into wood?

4. The screwdriver is one example of a type of simple machine called a wheel-and-axle. On a screwdriver,
   (a) where is the wheel?
   (b) where is the axle?

5. Name at least three other simple machines that you will use today, each of which is really a wheel-and-axle.

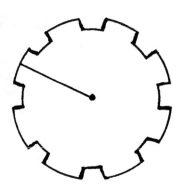

Figure 17

Using Simple Machines

We use simple machines like levers and wheels-and-axles for a variety of reasons:

(1) The machine may permit you to move a heavy load with very little effort.

(2) The machine may permit you to lift a load faster.

(3) The machine may permit you to apply your effort force from a more convenient direction.

SECTION 3 • Pulleys

Pulleys are very useful simple machines. You may find one or more in or around your home. A clothesline uses two pulleys. There may be pulleys in the system that allows you to open or close your window drapes.

Pulleys are often used to lift heavy loads. A single pulley like the one in **Figure 18** may be used to lift loads up to the top of a building. In this system, the effort force you exert is the same as the force of gravity on the load. This pulley is being used only because it makes it convenient to lift the load up from the ground to the top of the building. You can stand on the ground as you pull the load up, using the rope attached to the pulley.

Notice that a pulley is like a rotating lever. In **Figure 18**, if the pulley was a lever, the pivot would be in the middle, the load a certain distance to the left, and the effort the same distance to the right. (The lever arms are equal.) What class of lever would this be?

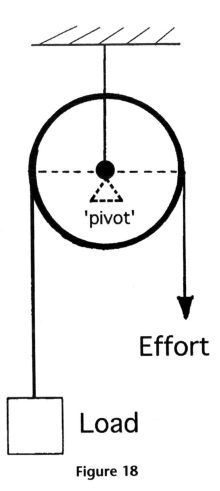

Figure 18

In **Figure 19**, a single pulley is being used in a different way, to lift the same load as the pulley in **Figure 18**.

Questions

1. If this pulley was a lever, where would the pivot be?

2. Where is the load?

3. Where is the effort applied?

4. Notice that the lever arm for the effort is twice as long as the lever arm for the load! How would the effort force compare with the load?

5. What class of lever would this be?

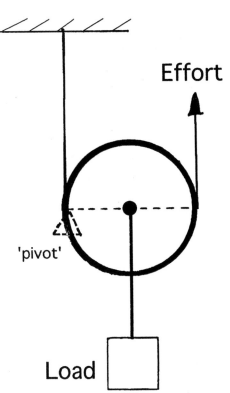

Figure 19

3.1 Try This Yourself! Which Person will Move?

The Problem

There are two stools about two metres apart, on a smooth floor. The person sitting on the first stool is heavier than the person on the second stool (see **Figure 20**). A single clothesline pulley is attached to the heavier person's stool. A rope is tied to the lighter person's stool, and then runs through the pulley, as in the diagram.

Figure 20

Make a prediction!

PULL SLOWLY AND STEADILY ON THE ROPE TO AVOID TOPPLING THE STOOLS

If someone pulls on the free end of the rope, which of the two people will slide faster along the floor? Will it be the heavier person on the left or the lighter person on the right?

What You Need

2 stools
1 clothesline pulley
1 long length of rope (about 3 m)
1 short length of rope (about 50 cm)

What to Do

After you make your prediction, set up the situation shown in **Figure 20**. Have someone pull on the rope with a steady force. What happens?

Questions

1. Which person moved, the lighter one or the heavier one?

2. Try to explain the results of this experiment.

Using a System of Simple Machines

Many devices we use every day do not consist of one simple machine, but of a combination or **system** of two or more simple machines. In the next activity, you will use two simple machines at once to make the job of lifting a heavy load easier. The load in this situation is a toy cart.

3.2 *Try This Yourself! Making Things Easier*

What You Need

1 toy cart
1 spring balance
1 single pulley
1 ramp (a bookshelf)

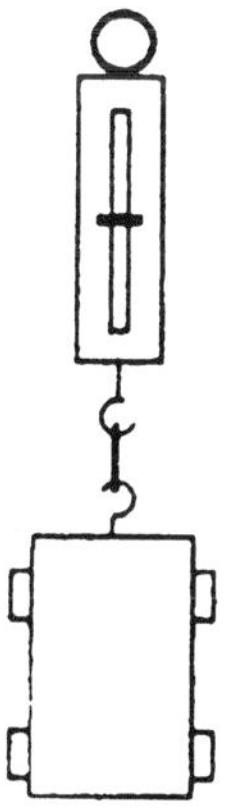

What to Do

1. Use a spring scale to measure the force needed to lift the toy cart straight up (see **Figure 21**).

   Effort Force = Load = <u>? newtons</u>

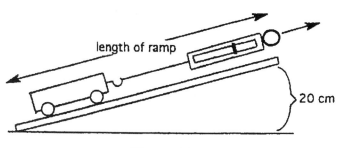

Figure 21

2. Place a 'ramp' (loose book-shelf) on your desk, and prop one end up so that it is about 20 cm above the surface of your desk (see **Figure 22**).

3. Attach your spring balance to the toy cart again, and this time measure the force you need to pull the toy cart up the ramp.

   Effort Force = <u>? newtons</u>

length of ramp

20 cm

Figure 22

Note! The ramp is a simple machine, usually called an **inclined plane**.

Question

Does the ramp reduce the effort force you need to lift the toy cart?

4. Attach a single pulley to the cart, as in **Figure 23**. Tie one end of a string to the ring of a tall ring stand (or some other support). Pass the string through the pulley, and attach a spring scale to the free end of the string. Lift the cart *straight up* using the pulley, and measure the force you need to do this.

   Effort Force = <u>? newtons</u>

Question

Does the pulley reduce the effort you need to lift the toy cart straight up?

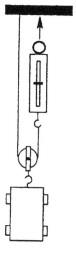

Figure 23

5. Combine the pulley and the ramp, as in **Figure 24**. Measure the effort force you need to pull the toy car up the ramp when you use both simple machines together as a system.

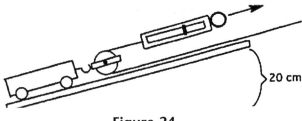

20 cm

Figure 24

Effort Force = <u>? newtons</u>

Questions

1. Compare the force you had to exert to lift the toy cart straight up (the actual **load**), with the **effort force** you needed to pull it up the ramp, using the pulley. (Divide the load by the **effort force**.)

2. What is the **mechanical advantage** of your system?

3. If you lifted the toy cart directly up, you would only have to move it a distance of 20 cm (the height of the ramp). When you pulled the cart up the ramp using the two simple machines, was the distance over which you exerted the effort about the same, far less, or far more?

Challenge!

Measure how far you have to pull on the string to raise the load 20 cm. This distance if the **effort distance**. Compare the **effort distance** with the **load distance**, the same way you compared the load with the effort force. (Divide the **effort distance** by the **load distance**.)

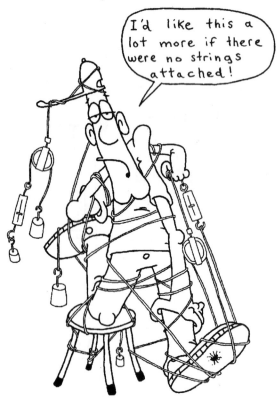

SECTION 4 • Other Simple Machines

So far, you have worked with four kinds of simple machine: the lever, the wheel-and-axle, the ramp (inclined plane) and the pulley. There are several other kinds of simple machine. You use all of them regularly, perhaps without even thinking about it.

The Wedge

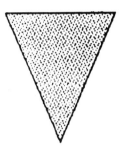

Figure 25

A **wedge (Figure 25)** can take many forms. Axes and knives are wedges. Chisels, saw teeth and needles are also wedges. Golf clubs are wedges. (One of them is even called a 'wedge'!)

Your teeth act like wedges. The sharper a wedge is, the more it will multiply your effort force.

Figure 26

Figure 27

Figure 26 illustrates an axe, which is a type of wedge, and an actual wedge. Both of these simple machines make it easier to split wood for a fireplace. **Figure 27** is a golf 'wedge'. This particular wedge is used to hit shots out of sand traps, so it is called a sand wedge.

The Screw

Wood screws, bolts, vises and propellers are all **screws**. **Figure 28** illustrates a typical wood screw.

EXERCISE CARE
WHEN USING
SHARP EQUIPMENT

Figure 28

Figure 29

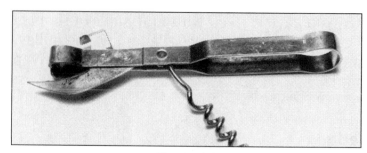

Figure 30

In **Figure 29**, a wood screw is being driven into a block of wood. **Figure 30** is a can opener, with a corkscrew attached. How many simple machines can you see in this can opener?

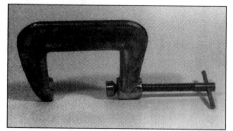

Figure 31

Figure 32

Exercise Care When Using Sharp Equipment

Figure 31 is a C-clamp. It has a screw and a lever. **Figure 32** is a nutcracker. It has both a screw and a wheel-and-axle.

Figure 33

Figure 34

Figure 33 is a photograph of a cork remover for a wine bottle. Examine the photograph closely. Can you find the screw? A lever? A wheel-and-axle? Gears are also simple machines. Can you see the gears? **Figure 34** is a photograph of a propeller on a motorboat. Aircraft propellers (**Figure 35**) are also screw-type simple machines.

Figure 35

The Lift

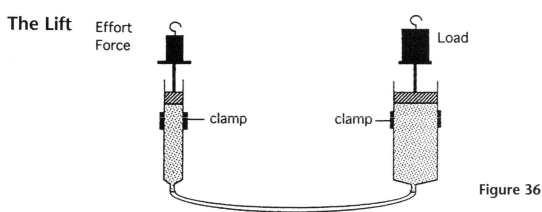

Figure 36

4.1 Try This Yourself! A Simple Machine That Uses Fluids

What To Do

1. Arrange two syringes as in **Figure 36**. They are joined by plastic tubing. One syringe is larger than the other. Clamp both syringes vertically on ring stands, if possible.
2. Place a heavy mass on the platform of the larger cylinder. Push down gently on the platform of the smaller cylinder.
3. Measure how far you have to push down on the piston of the smaller syringe **(effort distance)** in order to move the load on the larger syringe up a distance of 5 mm.
4. Use your model lift to raise a heavy object (like a large book).

The arrangement you have just used is a model of a **pneumatic lift**. (Pneumatic refers to air pressure.) When a liquid is used inside the lift, it is an **hydraulic lift**.

Question

Make a list of at least six examples of equipment used in industry that use the same principle as the hydraulic lift (one example is automobile brakes).

SECTION 5 • Gears

What Is A Gear?

A **gear** is a toothed wheel, connected to a shaft, which usually interlocks, or **meshes**, with another toothed wheel.

Two or more meshed gears form a **gear train**. A gear train is a type of simple machine.

Gear trains may be used:

(1) to transfer a force from one part of a machine to another;
(2) to move a heavy load using a small effort force;
(3) to slow down a rotary motion;
(4) to speed up a rotary motion; or
(5) to change the direction of a rotary motion.

Driving Gears and Driven Gears

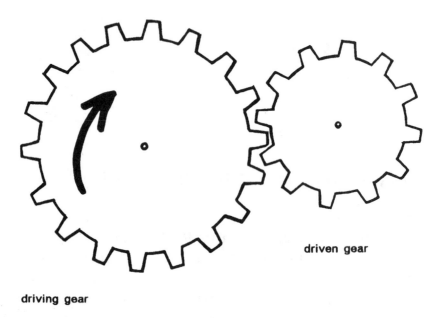

driven gear

driving gear

Figure 37

In a gear train, the gear that supplies the driving force is called the **driving gear**. The gear to which the force is applied is the **driven gear**. See **Figure 37**. If two meshing gear wheels in a gear train are of different sizes, the smaller of the two is called the **pinion**.

Gear Ratios

For any pair of meshed gear wheels (**Figure 37**), a useful number is the ratio you obtain when you divide the number of teeth on the larger gear wheel (gear) by the number of teeth on the smaller gear wheel (pinion). This number is called the **gear ratio**. For example, in **Figure 37**, the gear has 18 teeth and the pinion has 12 teeth. The gear ratio is therefore:

$$\frac{\text{Number of Teeth on Gear}}{\text{Number of Teeth on Pinion}} = \frac{18}{12} = \frac{3}{2}.$$

You would say, "The gear ratio is 3 to 2."

5.1 Try This Yourself! How Do Gears Change Rotation Speed?

Part A   How do gears *speed up* rotary motion?

Figure 38

1. Set up the arrangement in **Figure 38**, using a gear kit.

2. Which is the **pinion** in this arrangement, the **driving gear** or the **driven gear**?

3. Turn the **driving gear** (left) in a clockwise [↗↘] direction. Does the **driven gear** move clockwise [↗↘] or counterclockwise [↙↖]?

4. How many teeth are there (a) on the **driving gear**? (b) on the **driven gear**?

5. What is the **gear ratio** for this gear train?

6. Predict how many complete turns the **driven gear** will make for every one turn of the **driving gear**.

7. (a) Place a small removable mark on one tooth of each gear wheel.
    (b) Rotate the **driving gear** one full turn. While you do this, count how many rotations the **driven gear** makes.
    (c) Does your count agree with your prediction?
    (d) How does your count compare with the **gear ratio** of this gear chain?

Part B   How do gears *slow down* rotary motion?

Figure 39

1. Set up the arrangement in **Figure 39**.

2. Which is the **pinion** in this arrangement, the **driving gear** or the **driven gear**?

3. Turn the **driving gear** (left) in a clockwise [↗↘] direction. Does the **driven gear** move clockwise [↗↘] or counterclockwise [↙↖]?

4. Predict how many turns the **driven gear** will complete when the **driving gear** is rotated one full turn. Test your prediction as you did in **Part A**.

5. Count how many turns the **driving gea**r must make in order to rotate the **driven gear** one full turn. How does this number compare with the **gear ratio** of this gear train?

Questions

1. Examine the hand-operated food mixer in **Figure 40**, and the hand-operated drill in **Figure 41**. Decide whether these devices *increase* or *decrease* rotational speed. The gears in these devices are clearly visible.

2. (a) Does the gear train in the winch in **Figure 42** *increase* rotational speed or *decrease* it?
   (b) Will the gear train in the winch *increase* or *decrease* the **effort force** you will need to pull the object to which the winch is attached ?

Figure 40

Figure 41

Figure 42

EXERCISE CARE
WHEN USING
SHARP EQUIPMENT

5.2 Try This Yourself! Using Gears to Change Direction of Rotation

Figure 43

1. Set up the gear train in **Figure 43**. Rotate the driving gear in a clockwise direction. In which direction does the driven gear rotate, clockwise or counterclockwise?

Figure 44

2. **(a)** Set up the gear train in **Figure 44**. An extra gear called the idler is inserted between the driving gear and the driven gear.
   (b) Rotate the driving gear in a clockwise direction. In what direction does the driven gear rotate? In what direction does the idler rotate?

Figure 45

3. **(a)** Set up the gear train in **Figure 45**. The driven gear being used here is called a **crown gear**, because of its crown-like shape. Notice that the two axles, or shafts, form an angle of 90° with each other.

   (b) Rotate the driving gear by using the handle. What happens to the driven gear? Where might you use such a gear train arrangement?

Question

For the gear train you used in **Procedure 3**, what was the **gear ratio**?

SECTION 6 • Test Your Knowledge!

Reviewing Simple Machines

(1-10) Answer the questions about the simple machine in each photograph.

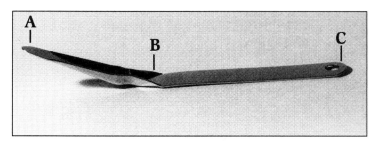

Figure 46

1. What **class of lever** is the staple remover in **Figure 46**?
   A. First   B. Second   C. Third

2. Where is the **load** in this simple machine?
   (A, B, C)

3. Where is the **fulcrum**?
   (A, B, C)

4. Where is the **effort force** applied?
   (A, B, C)

5. What **class of lever** is the hole punch in **Figure 47**?
   A. First   B. Second   C. Third

6. Where is the **load** in this simple machine?
   (A, B, C)

7. Where is the **fulcrum**?
   (A, B, C)

8. Where is the **effort force** applied?
   (A, B, C)

9. What **type of simple machine** is shown in **Figure 48**?
   A. wheel-and-axle
   B. lever
   C. pulley

10. Where is the effort force applied?
    (A, B, C, D)

EXERCISE CARE
WHEN USING
SHARP EQUIPMENT

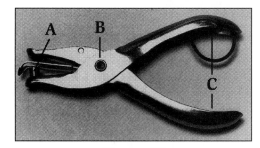

Figure 47

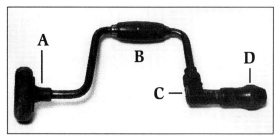

Figure 48

11. A winding pathway up a mountainside is an example of
    A. a lever
    B. a wheel-and-axle
    C. an inclined plane
    D. a pulley

12. What two kinds of simple machine are there in a pair of scissors?

13. What two kinds of simple machine do you use when you chew gum?

14. What is the gear ratio for each of the gear chains shown in **Figures 49 (a)**, **(b)** and **(c)**?

    (Divide the number of teeth on the gear by the number of teeth on the pinion.)

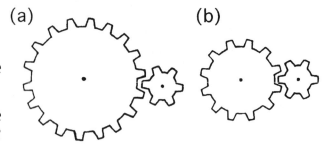

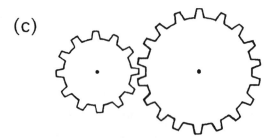

Figure 49

15. **Figure 50** shows the gears inside an electric food mixer. An electric motor rotates the driving gear in the centre, which is called a **worm gear**.
    (a) If the driven gear on the left rotates *clockwise*, in what direction will the driven gear on the right rotate?
    (b) Does this gear chain increase **force** or increase **rotational speed**?
    (c) What other reason is there for using this arrangement of gears?

Figure 50

EXERCISE CARE
WHEN USING
SHARP EQUIPMENT

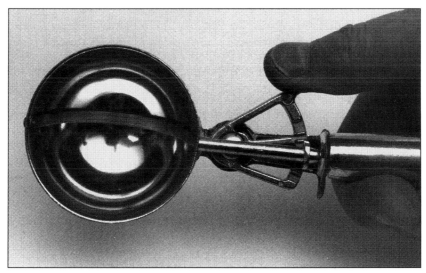

Figure 51

16. Examine an ice cream scoop like the one shown in **Figure 51**.
    (a) Describe how the gears on the scoop work.
    (b) Does the gear chain used in the ice cream scoop increase **force** or **rotational speed**?

Figure 52

17. Examine the gear system used on an electric drill to tighten the drill bit into position (see **Figure 52**).
    (a) Where is the driving gear?
    (b) Does this gear chain increase force or increase rotational speed?

EXERCISE CARE
WHEN USING
SHARP EQUIPMENT

Project

Examine a mountain bike closely.

1. Find as many examples of simple machines on the bike as you can, and list them.

2. How many 'speeds' is this bike? How are the gears used to achieve the different 'speeds'?

3. What is the difference between 'low gear' and 'high gear'?

4. How does a mountain bike rider use gears when climbing a steep hill?

5. How does a mountain bike rider use gears when travelling full speed on a level track?

Show What You Have Learned!

1. Use scrap materials and anything your teacher might supply to build a system of simple machines that can lift a load with an effort force of less than one-fifth of the load. (Its mechanical advantage will be at least five.)

2. Design a system made up of no more than three single pulleys that can lift a load with an effort force of no more than one-eigth the load. (Its mechanical advantage is at least eight.)

3. Design a simple machine that performs a useful task and provides a large speed advantage. (The mechanical advantage will be less than one.)

4. Design a system of several simple machines that accomplishes some very simple task, but, in the interests of humour, involves a ridiculously long chain of events in order for the task to be accomplished.

   Hint! Find out who **Rube Goldberg** was, and look for examples of his famous 'Rube Goldberg' inventions. Then, see if you can create a working model of one of your own 'inventions'.

Other Science and Technology titles from Trifolium Books

SPRINGBOARDS FOR TEACHING SERIES

INVENTEERING: A Problem-Solving Approach to Teaching Technology

Bob Corney & Norm Dale

An essential "getting started" resource for teachers of **Grades 1–8** wanting to provide their students with hands-on technological experiences.

8½" x 11" • 128 pages • Soft cover • Illustrations
ISBN: 1-55244-014-1 • $29.95 Can. • **Available**

IMAGINEERING: A "Yes, We Can!" Sourcebook for Early Technology Experiences

Bill Reynolds, Bob Corney, and Norm Dale

Packed with ideas to stimulate young students' imagination and creativity as they explore the issues and applications of technology. For teachers of **Grades K–3.**

8½" x 11" • 144 pages • Soft cover • Illustrations
ISBN: 1-895579-19-8 • $29.95 Can. • **Available**

ALL ABOARD!: Cross Curricular Design and Technology Strategies and Activities

By Metropolitan Toronto School Board teachers

This teacher-tested resource helps educators integrate design and technology easily and effectively into day-to-day lessons. For teachers of **Grades K–6.**

8½" x 11" • 176 pages • Soft cover • Illustrations
ISBN: 1-895579-86-4 • $21.95 Can. • **Available**

Take a Technowalk to Learn about Materials and Structures

Peter Williams & Saryl Jacobson

Provides teachers of **Grades K–8** with 10 fun Technowalks designed to encourage students to investigate the **materials and structures** that surround us.

8½" x 11" • 96 pages • Soft cover • Illustrations
ISBN: 1-895579-76-7 • $21.95 Can. • **Available**

Take a Technowalk to Learn about Mechanisms and Energy

Peter Williams & Saryl Jacobson

Now teachers of **Grades K–8** have fun with 10 new Technowalks designed to encourage students to investigate **mechanisms and energy** in the classroom, school and community.

8½" x 11" • 92 pages • Soft cover • Illustrations
ISBN: 1-55244-004-4 • $25.95 Can. • **Available**

TEACHERS HELPING TEACHERS SERIES

BY DESIGN
Technology Exploration and Integration

By the Metropolitan Toronto School Board Teachers

Over 40 open-ended activities for **Grades 6–9** integrate technology with other subject areas.

8½" x 11" • 176 pages • Soft cover • Illustrations
ISBN: 1-895579-78-3 • $39.95 Can. • **Available**

Mathematics, Science, & Technology Connections

By Peel Board of Education Teachers

Twenty-four exciting integrated Math, Science, and Technology activities for **Grades 6–9.**

8½" x 11" • 160 pages • Soft cover • Illustrations
ISBN: 1-895579-37-6 • $39.95 Can. • **Available**